YOUR KNOWLEDGE HAS VALUE

- We will publish your bachelor's and
  master's thesis, essays and papers

- Your own eBook and book -
  sold worldwide in all relevant shops

- Earn money with each sale

Upload your text at www.GRIN.com
and publish for free

Bibliographic information published by the German National Library:

The German National Library lists this publication in the National Bibliography; detailed bibliographic data are available on the Internet at http://dnb.dnb.de .

Imprint:

Copyright © 2017 GRIN Verlag, Open Publishing GmbH
Print and binding: Books on Demand GmbH, Norderstedt Germany
ISBN: 9783668524026

This book at GRIN:

http://www.grin.com/en/e-book/373936/bligh-s-lane-s-theory-of-seepage

Florante Jr Poso

Bligh's & Lane's Theory of Seepage

GRIN Publishing

Florante C. Poso, Jr., MCE, PhD

2017

<u>TABLE OF CONTENT</u>

I – Bligh's Creep Theory

II – Lane's Weighted Creep Theory

1.1 - Bligh's Creep Theory

To prevent internal erosion and particle migration, control of seepage pressures and velocities must be given due consideration in the design of hydraulic structures.

The percolation length (seepage) for a foundation can be determined by using various methods. There are number of methods available to analyze the problem on seepage and uplift pressure, and one of which is Bligh's theory of creep. Other methods are Lane's Method, Kosla's Theory and Flow nets.

Based on Bligh's theory, that along the bottom contour of the structure, the water creeps, and the percolation length (seepage) can be determined.[1]

1.2 Concept of the Theory:

The water which percolates into the foundation creeps through the joint between the profile of the base of structure and the subsoil.[2]

The seeping water comes out at the downstream end. Then, water travels along the vertical, horizontal or inclined path without making any distinction.[3]

The head of water lost in the path of percolation is the difference of water levels on the upstream and the downstream ends. The imaginary line which joins the water levels on the upstream and the downstream end is called a hydraulic gradient line.[4] (*Refer to Figure 1*)

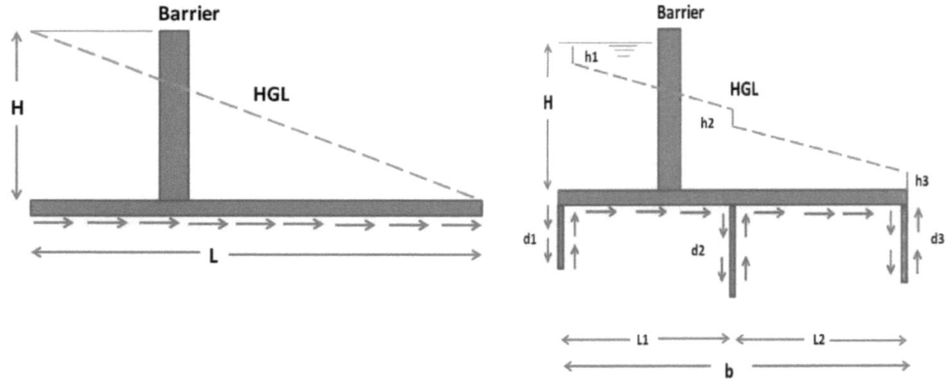

Figure 1: Illustration of Bligh's creep theory (Patterned from: Santosh Kumar Garg, Irrigation Engineering and Hydraulic Structures, 2006)

---

[1] Santosh Kumar Garg, Irrigation Engineering and Hydraulic Structures, Khanna Publishers, Delhi:2006, p 553

[2] **Shreyasi Sen, "Bligh's Creep Theory for Design of Weir on Permeable Foundation", Your Article Library, http://www.yourarticlelibrary.com/water/water-engineering/blighs-creep-theory-for-design-of-weir-on-permeable-foundation/61166/**

[3] Ibid

[4] "Bligh's Creep Theory for Seepage Flow", The Constructions of Deductions, 2015, https://yamannvinci069.blogspot.com/2015/08/blighs-creep-theory-for-seepage-flow.html

<u>**1.3 Summary of Bligh's assumptions:**</u>[5]

- ✓ The percolating water creeps along the base profile of the structure, which is in contact with the subsoil.
- ✓ <u>Line of creep</u> is the path of percolation along the base of the structure.
- ✓ <u>Creep length</u> is the total length of the path traversed by the percolating water. Total length covered by the percolating water till it emerges out at the downstream end.
- ✓ The length of the line of creep includes the vertical distances along both sides of cutoff walls or curtain of sheet piling (if any), as well as horizontal distance along the apron.
- ✓ The head loss per unit length of creep (hydraulic gradient) is proportional to the distance of the point from the upstream of the foundation (constant).

The limitation of this theory is that it does not differentiate between the horizontal and vertical creeps in estimating the exit hydraulic gradient.

The Creep length, L[6]

$$\boxed{L = b + 2d_1 + 2d_2 + 2d_3}$$

The hydraulic gradient or the loss of head per unit length of creep is,

$$\frac{H}{L} = \frac{H}{b + 2d_1 + 2d_2 + 2d_3}$$

Refer to Figure 1:

where:

b = the total horizontal distance
$d1$; $d2$; $d3$ are length of piles
H = total height of water upstream

Based on the assumptions, the following conclusions are derived:

- For any point, the head loss is proportional to the creep length.
- As the hydraulic gradient is constant, if L_1 is the creep length up to any point, then head loss up to any point will be $(H/L) L_1$ and the residual head at this point will be $(H - (H/L) L_1)$.
- Considering the cutoffs, the head losses will be:
  $(H/L) 2d_1$, $(H/L) 2d_2$ and $(H/L) 2d_3$
- The reciprocal of the hydraulic gradient (L/H) is known as Bligh's coefficient of creep, $C = L/H$.

---

[5] "Bligh's Creep Theory", http://www.aboutcivil.org/bligh%27s-creep-theory-of-hydraulic-structures.html
[6] Santosh Kumar Garg, Irrigation Engineering and Hydraulic Structures, Khanna Publishers, Delhi:2006, p 554

<u>**1.4 Safety against piping:**</u>[7]

In order to ensure that the structure is safe againt piping, the following should be taken into consideration:

- The creep length should be sufficient to provide a safe hydraulic gradient according to the type of soil (L = CH)

- Bligh recommended certain values of C for different soils presented in Table 1.

- The hydraulic gradient (H/L) should be equated to 1/C
  H/L = 1/C (for the soil)

Table 1: Bligh coefficient of creep C and safe hydraulic gradient

Type of soil	Value of C	Safe Hydraulic Gradient
Light sand &mud (River Nile)	18	1/18
Fine Micaceous sand	15	1/15
Coarse grained sand	12	1/12
Sand mixed with boulder and gravel; and for loam soil	5 to 9	1/9 to1/5
Gravel	5	1/5

Source: Shreyasi Sen, "Bligh's Creep Theory for Design of Weir on Permeable Foundation", Your Article Library, http://www.yourarticlelibrary.com/water/water-engineering/blighs-creep-theory-for-design-of-weir-on-permeable-foundation/61166/

<u>**1.5 Safety against uplift pressure:**</u>[8]

- The uplift pressure (residual seepage head) at that point is the ordinate of the subsoil hydraulic gradient line above the bottoms of the floor at any point.

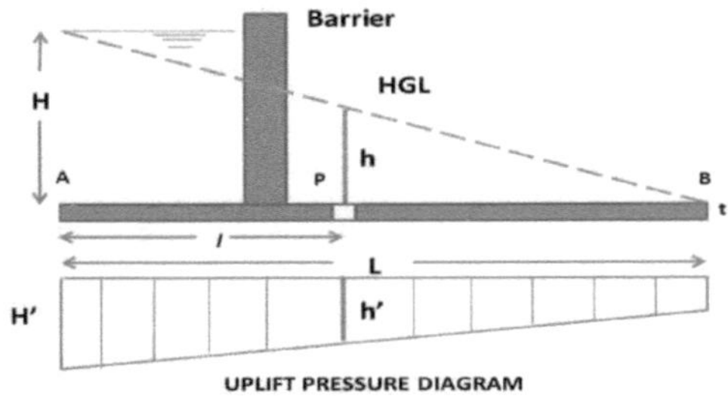

Figure 2. Uplift Pressure Diagram (Source: Author's own work)

---

[7] Ibid

[8] M. R. Kabir, ECONOMICAL & PHYSICAL JUSTIFICATION FOR CANAL, Chapter 5, http://www.uap-bd.edu/ce/Handouts/CE-461/Doc/Chapter-5.pdf

If h' is the uplift pressure head at a point under the floor, the pressure intensity is,[9]

$$P = \rho g h'$$

This pressure is to be resisted by the weight of the floor with thickness (t) and density ρ_m (for concrete = 2400 kg/m^3).

The downward force per unit area due to the weight of the floor is:

$$W = \rho_m g\, t$$

Therefore, equating

$$\rho_m g\, t = \rho g h'$$

which gives:

$$h' = \frac{\rho_m t}{\rho} = S_m t$$

Where:

S_m is the relative density of the floor material. Thus, we can write,

$$h' - t = S_m t - t$$

and for the thickness of the floor, [10]

$$t = \frac{h' - t}{S_m - 1} = \frac{h}{S_m - 1}$$

where h is the pressure head (ordinate of hydraulic gradient) measured above the top of floor, and

(S_m-1) is submerged specific gravity of the floor material.

A safety factor of 4/3 to 3/2 can be applied in the design of thickness:

$$t = \frac{4}{3} \frac{h}{S_m - 1} \quad to \quad \frac{3}{2} \frac{h}{S_m - 1}$$

Assuming a value of   Sm= 2.24,

Then, $t \approx 1.08\,h$ to $1.2\,h$

[9] Santosh Kumar Garg, Irrigation Engineering and Hydraulic Structures, Khanna Publishers, Delhi:2006, p 555
[10] "Bligh's Creep Theory for Seepage Flow", The Constructions of Deductions, 2015,
https://yamannvinci069.blogspot.com/2015/08/blighs-creep-theory-for-seepage-flow.html

For design efficiency, the following notes should be taken into consideration:

- The design will be economical if the greater part of the creep length (impervious floor) is provided upstream of the weir where nominal floor thickness would be sufficient.
- The downstream floor has to be thicker to resist the uplift pressure. However, a minimum floor length is always required to be provided on the downstream side from the consideration of surface flow to resist the action of fast flowing water whenever it is passed to the downstream side of the weir.
- The provision of maximum creep length on the upstream side of the barrier structure also reduces uplift pressures on the portion of the floor provided on the downstream side of the barrier.
- A vertical cutoff at the upstream end of the floor reduces uplift all over the floor.
- According to Bligh's theory, a vertical cutoff at the upstream end of the floor is more useful than the one at the downstream end of the floor.

1.6 Limitations of Bligh's Theory[11] [12] [13]

- Bligh made no distinction between horizontal and vertical creep.
- The theory holds good as long as horizontal distance between cut-offs or pile lines is greater than twice their depth.
- No distinction is made between the effectiveness of the outer and inner faces of sheet piles and short and long intermediate piles.
- Later investigations have shown that the outer faces of the end piles are much more effective than the inner ones.
- Intermediate piles of shorter length than the outer ones are ineffective except for local redistribution of pressure.
- No indication on the significance of exit gradient.
- Average value of hydraulic gradient gives idea about safety against piping.
- For safety purposes, the exit gradient must be less than critical exit gradient.
- The loss of head is proportional to creep length (assumption) is not true and actual uplift pressure distribution is not linear, but it follows a sine curve.
- Bligh did not specify the absolute necessity of providing a cut-off at the downstream end of the floor, whereas it is absolutely essential to provide a deep vertical cutoff at the downstream end of the floor to prevent undermining.

---

[11] Shreyasi Sen, "Bligh's Creep Theory for Design of Weir on Permeable Foundation", Your Article Library, http://www.yourarticlelibrary.com/water/water-engineering/blighs-creep-theory-for-design-of-weir-on-permeable-foundation/61166/

[12] "Bligh's Creep Theory", http://www.aboutcivil.org/bligh%27s-creep-theory-of-hydraulic-structures.html

[13] [13] Fetene Nigussie, Hydraulic Structures II-Lecture Note, https://www.scribd.com/document/305701128/Chapter-4-PART-1

<u>**1.7 SAMPLE PROBLEMS:**</u>

Example 1:
Find the hydraulic gradient and uplift pressure at a point 12m from the upstream end of the floor as shown in the figure below.

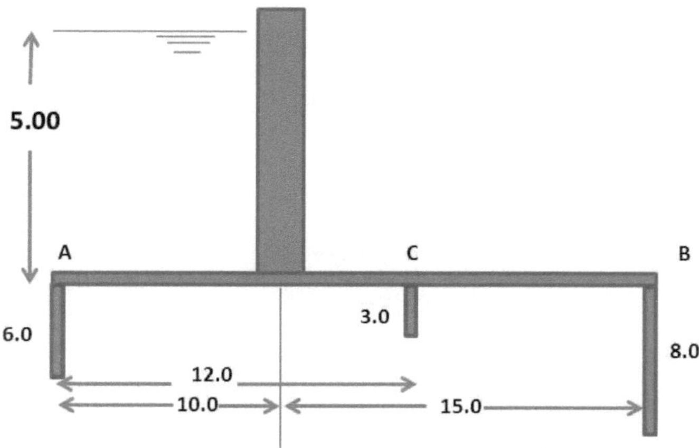

Source: Author's own work

Solution:

The water pecolates at point A and emerges at point B.

Total creep length
$$2(6) + 10 + 2(3) + 15 + 2(8) = 12 + 10 + 6 + 15 + 16 = \underline{\textbf{59m}}$$

Head of water on structure = 5m

Hydraulic gradient = 5/59 = 1/11.80

Type of soil	Value of C	Safe Hydraulic Gradient
Sand mixed with boulder and gravel; and for loam soil	5 to 9	1/9 to 1/5

From the table, the structure would be safe on sand mixed with boulders and gravel.

Calculate the Creep length at point C:

$L_1 = 2(6) + 2(3) + 12 = 12 + 6 + 12 = \underline{\textbf{30m}}$

$Hc = 5\,(29) / 59 = 2.46m$

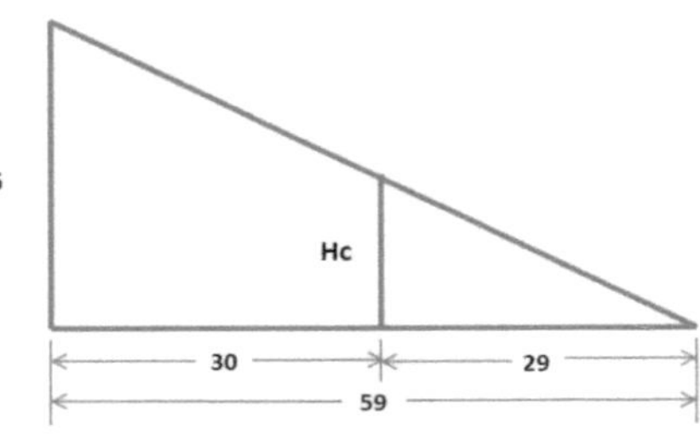

Source: Author's own work

Solving for the thickness at C, assume $S_m = 2.4$ (masonry)

$tc = Hc / (S_m\text{-}1) = 2.46 / (2.4 - 1.0)$

$tc = 1.76m$ of concrete

Apply factor of safety 4/3 to the formula

$\textbf{tc = 4/3 (1.76) = 2.35m}$

Example 2:

A hydraulic structure is built on fine sand as shown in Fig 1. Determine,
a) Whether the percolation gradient is safe if C for sand is 15.
b) Uplift pressure head at point A, B, C
c) Calculate the thickness of the floor at all the points A, B, & C.

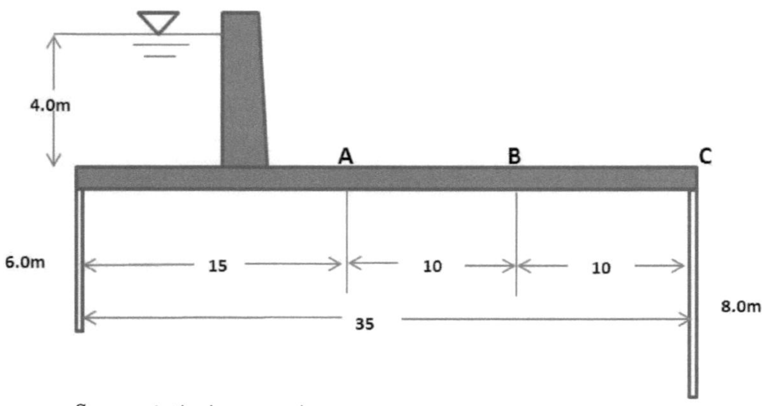

Source: Author's own work

a) Total head = H = 4m

Total creep length = L = 2(6) + 35 + 2(8) = 12 + 35 + 16 = **63m**

C = H/L = 4/63 = 1/15.75 less than 1/15, therefore it is safe!

No.	Type of soil	Value of C	Safe exit gradient less than
1	Fine sand	15	1/15

b) Calculate for creep lenths at Point A, B, & C (L1, L2, L3)

Let: L1 = 2(6) + 15 = 12 + 15 = 27m
        L2 = 2(6) + 15 + 10 = 12 + 15 +10 = 37m
        L2 = 2(6) + 15 + 10 + 10 = 12 + 15 + 10 +10 = 47m

Uplift Pressure at Point A
    H/L = Ha / (63-27)
    4/63 = Ha / 36
    Ha = 2.29 m

Uplift Pressure at Point B
    H/L = Hb / (63-37)
    4/63 = Ha / 26
    Ha = 1.65 m

Uplift Pressure at Point C
    H/L = Hc / (63-47)
    4/63 = Ha / 16
    Ha = 1.02 m

c) Thickness calculation, assume Sm for masonry as 2.4

$$tc = \frac{Hc}{Sm - 1}$$

Apply factor of safety 4/3 to the formula

Thickness at A = 4/3 (2.29/(2.4 -1) = 2.18m
Thickness at B = 4/3 (1.65/(2.4 -1) = 1.57m
Thickness at C = 4/3 (1.02/(2.4 -1) = 0.97m

<u>**2.1- Lane's Weighted Creep Theory**</u>

Lane's theory was patterned from the Bligh's creep theory but according to Lane, Bligh had only calculated the total length of creep by adding both the horizontal and vertical length of creep and part of its limitation is it does not make any distinction between the two creeps.[14]

Some experts had criticized Lane's method due to the fact that it is an empirical method and not based on any mathematical approach. However, the method is also widely used due to the simplicity on its approach.

<u>**2.2 Concept of the Theory:**</u>

From the analysis of 200 dams all over the world, Lane's concluded that horizontal creep is less effective in reducing uplift than vertical creep. Therefore, Lane's weighted creep theory suggested a factor of one-third (1/3) for horizontal creep against one (1) for vertical creep.[15]

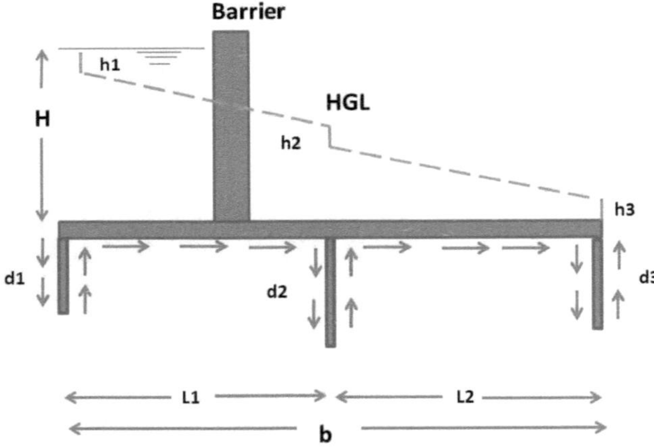

Figure 3. Length of creep illustration (Patterned from: Faculty of Engineering and Applied Science Hydraulic Structures CE404,University of Duhok, http://hyd.uod.ac/)

Application of the principle, the total length of creep will be:[16]

$$L = 2d1 + \frac{1}{3}L1 + 2d2 + \frac{1}{3}L2 + 2d3$$

$$L = \frac{1}{3}(L1 + L2) + 2(d1 + d2 + d3)$$

$$L = \frac{b}{3} + 2(d1 + d2 + d3)$$

---

[14] Lanes Weighted Creep Theory, The Construction of Deduction, 2015, https://yamannvinci069.blogspot.com/2015/08/lanes-weighted-creep-theory.html

[15] Santosh Kumar Garg, Irrigation Engineering and Hydraulic Structures, Khanna Publishers, Delhi:2006, p 555

[16] Ibid

$$L = \frac{\Sigma Horizontals}{3} + \Sigma Verticals$$

Where:

d_1, d_2, d_3 = length of piles
L_1, L_2, b = horizontal length

Thus, the weighted creep length, L_w, is given by

$$L_w = \frac{1}{3} N + V$$

Where:

N = sum of all the horizontal contacts and all the sloping contacts less than 45^0 to the horizontal.
V = sum of all the vertical contacts and all sloping contacts greater than 45^0 to the horizontal.

2.3 Safety Against Piping:

To ensure safety against piping, the Lw should be greater than C_1H [17]

$$\boxed{\mathbf{Lw > C_1H}}$$

Where:

H = Total seepage head (difference in water head between upstream and downstream)
C_1 = Lane's coefficient (empirical) of creep

Further if the hydraulic gradient $\left(\dfrac{H}{L_w}\right) \leq \left(\dfrac{1}{C_1}\right)$ safety against piping can be ensured.

Table 1 presents the values of Lane's Safe Hydraulic Gradient for different types of Soils.

Table 1: Values of Lane's Safe Hydraulic Gradient for different types of Soils

Type of Soil	Value of Lane's Coefficient, C1	Safe lane's Hydraulic gradient should be less than
Very fine sand or silt	8.5	1/8.5
Fine sand	7.0	1/7
Coarse sand	5.0	1/5
Gravel and sand	3.5 to 3.0	1/3.5 to 1/3
Boulders, gravels and sand	2.5 to 3.0	1/2.5 to 1/3
Clayey soil	3.0 to 1.6	1/3 to 1/1.6

Source: Irrigation Engineering and Hydraulic Structures, Garg (2006)

---

[17] Kabir, M. R., Economical & Physical Justification for Canal, Chapter 5, http://www.uap-bd.edu/ce/Handouts/CE-461/Doc/Chapter-5.pdf

2.4 SAMPLE PROBLEMS:

Example1: From the figure below, determine the seepage pressure at Point C by Lane's method. Consider the thickness of slab

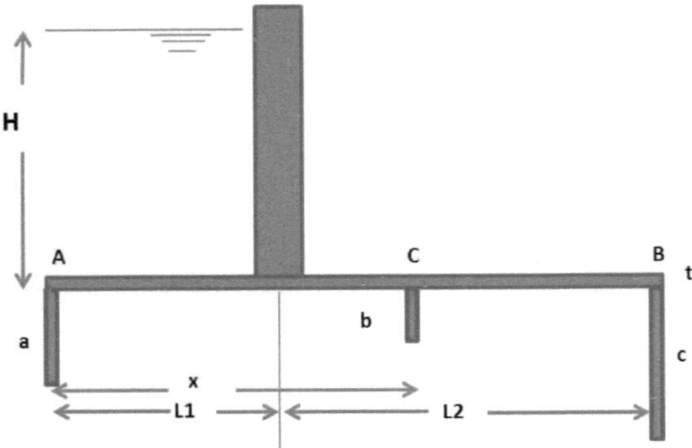

Source: Author's own work

ΣHorizontal $= L_1 + L_2$

ΣVertical $= t + 2a + 2b + 2c + t$

Total Length of creep (Lw) $= 1/3\ (L_1 + L_2) + (t + 2a + 2b + 2c + t)$

Horizontal distances from left to Pt. C $= x$

Vertical distance from left to Pt. C$= t + 2a + 2b$

Lx $= 1/3(L_1 + d) + (t_1 + 2a + 2b)$

$$hx = \frac{H\ (Lw - Lx)}{Lw}$$

Example 2. A hydraulic structure is built on fine sand (C = 15) as shown below. Using lane's weighted creep theory, determine:
   a) If the percolation gradient is safe
   b) Uplift pressure head at points A, B, C at a distance of 15, 25, and 35 m from the upstream end.
   c) Determine the thickness (t) of the floor at points A, B & C.
Assume C = 8.5,

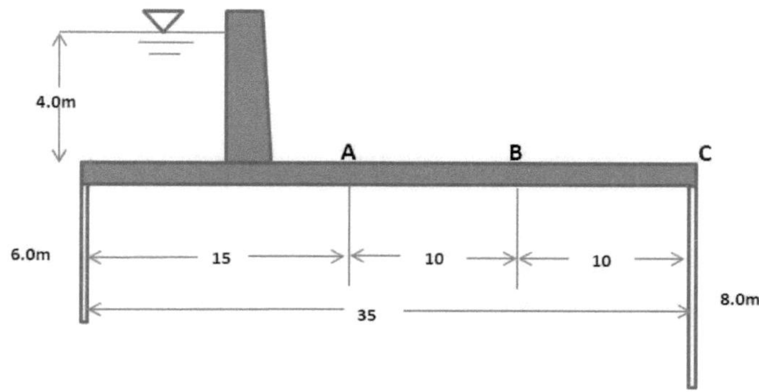

Source: Author's own work

a)

    Lw   = weighted creep length
           = 1/3 (15 + 10 + 10) + (2*6 + 2*8)
           = 1/3 (35) + (12 + 16) = 11.67 + 28
    Lw   = 39.67m

Hydraulic Gradient, = H/Lw = 4/39.67 = **1/9.92 (1/8.5 (SAFE!)**

a) Uplift pressures

$$L_a = 2\times6 + 1/3(15) = 12 + 5 = 17m$$
$$L_b = 2\times6 + 1/3(25) = 12 + 8.33 = 20.33m$$
$$L_c = 2\times6 + 1/3(35) = 12 + 11.67 = 23.67m$$

Pressure heads can be calculated using the formula:

$$h = H - \frac{H}{Lw}(La)$$

Where: H/Lw = 1/9.92

Pressure head at A:

$$H_a = 4 - \frac{17}{9.92} = 2.29m$$

Pressure head at B

$$H_b = 4 - \frac{20.33}{9.92} = 1.95m$$

Pressure head at C

$$H_c = 4 - \frac{23.67}{9.92} = 1.61m$$

Check:
Pressure head at C = 2*8 / 9.92 = 1.61 OK!

b) Thickness

For thickness;

$$t = \frac{4}{3}\frac{h}{S_m - 1} \quad to \quad \frac{3}{2}\frac{h}{S_m - 1}$$

Assume a factor of safety of 4/3 or 3/2, $S_m = 2.24$

$$\text{Thickness at A, } t = \frac{4}{3} \times \frac{2.29}{2.24 - 1} = 2.46m$$

$$\text{Thickness at B, } t = \frac{4}{3} \times \frac{1.95}{2.24 - 1} = 2.10m$$

$$\text{Thickness at C, } t = \frac{4}{3} \times \frac{1.61}{2.24 - 1} = 1.73m$$

Example 3.
1. Using the figure below, calculate the value of coefficient C & the uplift pressure at Point 2.

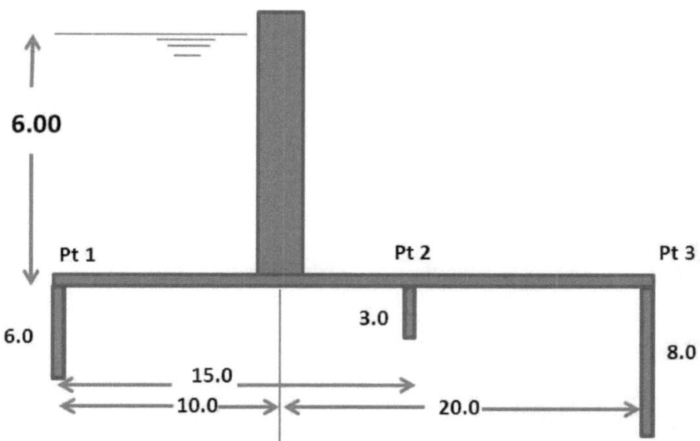

Source: Author's own work

Solution:

Σvert. $= (2)6 + 2(3) + 2(8) = 12 + 6 + 16 = 34$m

Σhor. $= 10 + 20 = 30$

Lw $= 1/3(\Sigma$hor. $) + \Sigma$vert. $= 1/3(30) + 34 = 10 + 34 = 44$m

Hydraulic Gradient, i

$i = $ H/Lw $\quad = 6/44 = $ __1/7.33 $\langle$ 1/8.5 for very fine sand or silt__

therefore, __C = 7.33__

Pressure head at Point 2

$L_2 = 2(6) + 2(3) + 1/3(15) = 12 + 6 + 5 = 23$m

$$H_2 = 6 - \frac{23}{7.33} = 2.86\text{m}$$

References:

Faculty of Engineering and Applied Science
Hydraulic Structures CE404
University of Duhok
http://hyd.uod.ac/

Bligh's Creep Theory for Design of Weir on Permeable
Foundationhttp://www.yourarticlelibrary.com/water/water-engineering/blighs-creep-theory-for-design-of-weir-on-permeable-foundation/61166/

Bligh's Creep Theory
http://www.aboutcivil.org/bligh%27s-creep-theory-of-hydraulic-structures.html

Fetene Nigussie, Hydraulic Structures II-Lecture Note,
https://www.scribd.com/document/305701128/Chapter-4-PART-1

Hydraulic Structures, 3rd Edition
P. Novak, A Moffat, C Nalluri and R. Narayanan
2001

Irrigation Engineering and Hydraulic Structures
Santosh Kumar Garg, 2006

Kabir, M. R., Economical & Physical Justification for Canal, Chapter 5, http://www.uap-bd.edu/ce/Handouts/CE-461/Doc/Chapter-5.pdf

Mohsin Siddique, Hydraulic Structures Chapter 4 Seepage Theories, University of Sharjah,
https://www.slideshare.net/yourmohsin/chapter-4-seepage-theories?from_action=save

The Constructions of Deduction
https://yamannvinci069.blogspot.com/2015/08/blighs-creep-theory-for-seepage-flow.html

List of Figures: